The Truth About SOLAR PANELS

The Book That Solar Manufacturers, Vendors, Installers And DIY Scammers Don't Want You To Read

Lacho Pop, MSE
&
Dimi Avram, MSE

Disclaimer Notice

The authors of this book, named 'The Truth About Solar Panels The Book That Solar Manufacturers, Vendors, Installers And DIY Scammers Don't Want You To Read', hereinafter referred to as the 'Book', make no representation or warranties with respect to the accuracy, applicability, fitness, or completeness of the contents of the Book. The information contained in the Book is strictly for educational purposes.

Summaries, strategies, tips and tricks are only recommendations by the authors, and the reading of the Book does not guarantee that reader's results shall exactly match the authors' results.

The authors of the Book have made all reasonable efforts to provide current and accurate information for the readers of the Book and the authors shall not be held liable for any unintentional errors or omissions that may be found.

The Book is not intended to replace or substitute any advice from a qualified technician, solar installer or any other professional and advisor, nor should it be construed, as legal or professional advice, and the authors explicitly disclaim any responsibilities for such use.

Installation of solar power systems requires certain background professional qualification and certification for working with high voltages and currents dangerous to human life, and for installing solar power systems and appliances. The reader should consult every step of the project or the installation with a qualified solar professional, installer or technician and local authorities.

The authors shall in no event be held liable to any party for any direct, indirect, punitive, special, incidental or other consequential damages arising directly or indirectly from any use of this Book, which is provided on "as is, where is" basis, and without warranties.

The use of this Book, provided by Digital Publishing Ltd should be based on reader's own due diligence, and the reader agrees that Digital Publishing Ltd shall not be liable for any success or failure.

This Book is ©copyrighted by Lachezar Popov and Digital Publishing Ltd and is protected under the applicable copyright laws and all other applicable intellectual property laws, with ALL RIGHTS RESERVED. No part of this BOOK may be copied, or changed in any format, sold, or used in any way other than what is outlined herein under any circumstances, without explicitly written permission of the authors.

About the Authors

Lacho Pop, MSE, has more than 15 years of experience in market research, technological research, design and implementation of various sophisticated electronic and communication systems. His large experience helps him present the complex world of solar energy in a manner that is both practical and easy-to-understand by a broad audience.

Dimi Avram, MSE, has more than 10 years of experience in the engineering of electrical and electronics equipment. He has specialized in testing electronic equipment and performing techno-economic evaluation of various kinds of electric systems. His excellent presentation skills help him explain even the most complex stuff to anybody interested.

You may contact the authors by visiting the website: solarpanelsvenue.com

Also by the authors:

The Ultimate Solar Power Design Guide: Less Theory More Practice [Kindle Edition]
ASIN: B00Q95UZU0

The New Simple And Practical Solar Component Guide [Kindle Edition]
ASIN: B00TR7IJPU

Top 30 Costly Mistakes Solar Newbies Make: Your Smart Guide to Solar Powered Home and Business [Kindle Edition]
ASIN: B00S8L4IIS

Contents

How can this book help you?

The book has been created to dispel misleading, rehashed or low-quality information about solar panels and solar power. The information out there is largely being provided by unqualified authors.

Buying, building or installing a solar electric (photovoltaic, PV) system for your home, business, vehicle or boat is a complex project with many nuanced and complex issues involved.

Whether you plan to do it yourself or hire a professional to do it for you, you need to know the correct terminology to research, design, purchase and communicate about your solar panel system.

Making the right selection of solar panels, however, will determine about 80% of the overall performance of your solar power system. Get it wrong and you will lose a lot of money and have to deal with associated problems for years to come.

This book is focused primarily on photovoltaic solar panels, how they perform and how they should be selected. It is written by solar experts especially for those who are just getting started.

This book contains no fillers and no fluff!

It gives you actionable and practical information about different types of solar panels, their features, and the different types of warranties available.

Moreover, the book honestly compares and reveals advantages and disadvantages of brand-new, secondhand and do-it-yourself solar panels in a way nobody else has done.

Armed with this valuable information, you can choose the best solar panel in terms of price, performance, and suitability for your solar project, while immediately spotting the "unfair" marketing tricks used to sell you an overpriced underperforming solar panel with a questionable warranty.

You can use the information provided in this book to judge the quality of solar quotes and filter out the disadvantageous ones.

Mind that most solar panel manufacturers, vendors, and installers often take advantage of the seeming complexity of solar parameters to market you an inferior solar panel in a better light and at a higher price, while at the same time keeping in line with consumer rights laws.

We hope that you have noticed the phrase 'seeming complexity'. By the end of this book, you are going to see that there is no complexity. What you are going to find is just concise, step-by-step, easy-to-read practical information written by real solar experts for those who are getting started.

Who is this book written for?

This book is for anyone who seeks practical information about solar panels, how they perform, and how to select solar panels for your solar project:

- Homeowners

- Do-it-yourself solar enthusiasts

- Lovers of recreational vehicles, campers, boats and other outdoor activities

- Survivalists

- Potential investors in solar power

- Business owners interested in solar power

- Students

- Teachers

- People interested in becoming solar installers

- People working in the sales and marketing area of solar power and green energy industry

- Many others keen on solar power and renewable energy.

Print this guide out to use as a reference while you are planning your PV system.

Write on it and take notes while you are surfing the web pricing out solar panels, and when you contact solar vendors and installers.

We wish you the best of luck with your solar project!

Basics of solar electrics

A solar photovoltaic (PV) system converts sunlight into electricity (electrical energy). Such a system can generate a part or all of your electricity demand.

A solar electric system can either reduce the amount of power you consume from your utility or entirely replace the utility grid.

The main component of every photovoltaic system is the photovoltaic (solar) module. A solar module consists of solar cells connected together. Further solar modules can be connected into solar arrays to achieve a higher energy yield.

In the picture on the following page, you can distinguish between the main types of photovoltaic units.

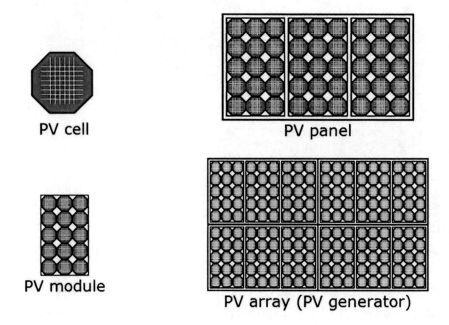

PV cell

PV panel

PV module

PV array (PV generator)

PV Cell -> PV Module -> PV Panel -> PV Array -> PV System

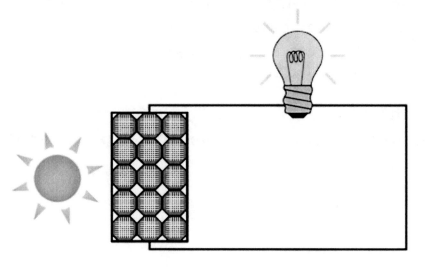

Photovoltaic cells and modules are made of semiconductor material (silicon) capable of producing electricity when exposed to sunlight. This phenomenon is called the 'photovoltaic effect'.

More energy is generated in sunny days, while less energy is generated in cloudy/rainy days, or when the photovoltaic array is shaded by obstructions (trees, lampposts, buildings, etc.).

Some types of solar cells and modules, however, are capable of providing more electricity in cloudy days, as they react more heavily to the ultraviolet component of the sun's spectrum, which is higher on such days.

The generated electrical energy can:

- Be used right away,

- Stored in a battery for later use, or

- Converted to AC electricity (most of the devices in your home or office operate on AC power), and then either used by in-home electrical appliances or exported to the utility grid.

Solar Panels

In this chapter, you will find:

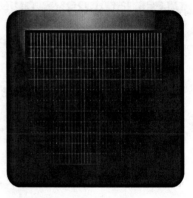

> *Important actionable and money saving information about different types of solar panels*

> *Lots of practical tips on how to choose the best solar panel type suitable for your solar project*

> *How to connect and how not to connect different solar panels*

Introduction

Nowadays the most widely used photovoltaic solar panels in solar power installations are:

- **Monocrystalline solar panels**

- **Polycrystalline solar panels**

- **Thin-film (amorphous) solar panels.**

Each solar panel type has its different:

- Price point

- Ability to convert solar energy into electricity, also known as efficiency

- Required installation area per generated DC electricity, also known as the necessary area for installation of 1 kWp (1kWp=1,000Wp)

These three main factors translate into two main key performance indicators used to compare different types of solar panels:

- $/Wp or price per installed Wp in USD

- The area needed for installation of 1 kWp.

Let's focus on the three important parameters of the various types of solar panels that determine the price point of the solar panel and the installation area needed for a solar array.

These three parameters are:

- Conversion efficiency

- Temperature's impact on the solar panel's efficiency

- The impact of the irradiance changes on efficiency – for example, the difference between solar performance on a sunny and on a cloudy day.

Conversion efficiency

Different types of solar panels have different abilities to convert sunlight into electricity. This is called 'conversion efficiency.'

Conversion efficiency according to solar panel type:

Monocrystalline Silicon	12.5-16%
Polycrystalline Silicon	11-14%
Copper Indium Gallium Selenide (CIGS)	10-18%
Cadmium Telluride (CdTe)	9-12%
Amorphous Silicon (a-Si)	5-7%

It should be noted that the conversion efficiency of a solar cell is usually higher than the efficiency of a solar panel.

According to Fraunhofer ISE: Photovoltaics Report (2014), updated 28 July 2014:

The record lab cell efficiency is 25% for monocrystalline and 20.4% for polycrystalline silicon wafer-based technology. The highest lab efficiency in thin-film technology is 19.8% for CIGS and 19.6% for CdTe solar cells.

Temperature's impact on solar panel efficiency

Another important key performance indicator of solar panels reveals how their ability to produce solar power efficiently is affected by temperature increase and hot environmental conditions.

This indicator is referred to as the **'temperature coefficient of power'**.

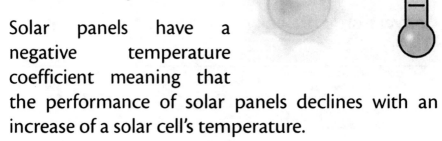

Solar panels have a negative temperature coefficient meaning that the performance of solar panels declines with an increase of a solar cell's temperature.

Solar panel rated output power is defined under Standard Test Conditions (STC), that is:

- 1,000 W/m2 of sunlight

- 25°C cell temperature

- Spectrum at air mass of 1.5.

Please be aware that, generally, for positive ambient temperatures, the temperature of a solar cell is about 15°C higher than the ambient temperature, as a result of solar panel encapsulation.

Example:

If the temperature coefficient of a solar panel is − 0.5%/°C and the ambient temperature is 40°C, the cell temperature is expected to be about 15°C higher than the ambient temperature because of panel encapsulation.

Therefore the loss in solar panel power output at 40°C will be:

0.5%/°C * (55°C − 25°C) = 15%.

In other words, a panel rated at 100W under STC (25°C cell temperature) will only produce 85W at 40°C ambient temperature.

So the temperature coefficient of thin-film solar panels is less negative compared to the temperature coefficient of crystalline panels.

Therefore, thin-film solar panels might produce more power than crystalline panels at elevated temperatures.

Impact of irradiance changes on solar panel efficiency

Solar irradiance is related to climate conditions. Solar irradiance changes throughout the day and the year, and is different in different places.

Therefore, the second important factor to consider when choosing the best solar panel type for your situation is the prevailing climate conditions at your location.

As we already know, performance parameters of solar panels are given under STC conditions, and any deviation from those conditions determines the amount of the pertaining losses.

The definition of STC conditions reveals the main factor contributing to the production losses:

- Temperature conditions

- Any deviation in respect to irradiance level of 1,000W/m2, corresponding to a common sunny day.

We've already described above the influence temperature has on the conversion efficiency of solar panels.

One of the most overlooked parameters is the influence of irradiance levels on solar panel productivity. Usually, the lower the level of irradiance with respect to 1,000W/m2, the lower the solar cell efficiency.

You can expect lower irradiance levels early in the morning, on a cloudy day or during winter.

Surprisingly, amorphous silicon thin-film solar modules, which are considered a lower grade and a cheaper version of the mono- and polycrystalline solar panels, demonstrate better performance in lower irradiance conditions than more expensive crystalline solar panels.

Therefore, the solar panel type that could be the best fit for your solar project should always be carefully chosen by weighing the pros against the cons of the aforementioned factors and parameters.

Solar panel types

Monocrystalline panels

Monocrystalline panels are the most efficient but also the most expensive. These panels can be found in blue or black color.

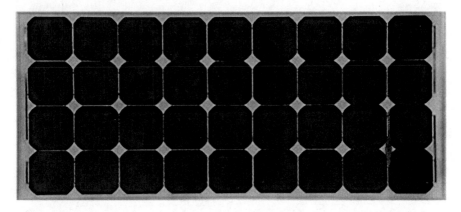

The fewer solar panels you need to produce a certain amount of power, the higher the efficiency. Normally, if there is not enough free space on your roof, you choose panels with higher efficiency. Monocrystalline panels are the oldest type of solar panels, are backed by a strong track record and are now considered a proven technology. Many monocrystalline panels that were installed more than 40 years ago are still perfectly operational.

Polycrystalline panels

Polycrystalline panels are slightly less efficient and cost 30-50% less than monocrystalline ones when intended to produce the same amount of solar generated power.

"Polycristalline-silicon-wafer 20060626 568" by Georg Slickers – Own work. Licensed under Creative Commons Attribution-Share Alike 3.0 via Wikimedia Commons – http://commons.wikimedia.org/wiki/File:Polycristalline-silicon-wafer_20060626_568.jpg

Polycrystalline panels have a lifecycle of about 25 years. Practice has shown, nevertheless, that polycrystalline panels installed more than 25 years prior are still perfectly operational.

Polycrystalline panels are typically blue in color and can be easily distinguished by their multifaceted, slightly-shimmering appearance.

Thin-film (amorphous) panels

Thin-film panels are the least expensive panels with the lowest efficiency – usually half the efficiency of monocrystalline panels. This means that, in order generate the same amount of power, you would need double the number of thin-film panels compared to monocrystalline ones.

Thin-film panels have a dark surface – usually brown, gray or black. Thin-film modules are commonly used in solar calculators.

The group of thin-film panels comprises the following subtypes:

- Amorphous-Si (a-Si)

- Tandem a-Si/microcrystalline

- CIGS (Copper Indium Gallium Selenide)

- CdTe (Cadmium Telluride)

- Dye-sensitized (TiO2)

The efficiency of amorphous silicon is about 6-7%, while the efficiency of CIGS thin-film panels is about 16-18%, with a recently achieved record of 20.8% in laboratory conditions.

Despite their high efficiency, CIGS thin-film panels fail to compete in terms of the lowest cost per produced Watt of electricity.

Another important issue worth mentioning is that thin-film panels are affected by the so-called 'Staebler-Wronski effect.'

This effect causes a reduction of module efficiency over time. The main reason that this occurs is the defect density of amorphous silicon, which tends to increase upon sunlight exposure.

About six months after installation this effect reaches an equilibrium and typically does not cause any further degradation of the solar module's output power. Therefore, you should mind that during the first year of operation, thin-film modules produce about 10-15% higher energy than standard.

After about six months of operation, however, they settle down to their usual yield, which will remain consistent over the remaining years.

The main drawback of the Stable-Wronski effect is that it requires the solar power system based on thin-film amorphous solar panels to be a little oversized, in order to sustain the higher initial output. Such an oversizing increases the cost of the solar panel system. Usually, the rated power of thin-film panels is given with respect to reaching a stabilized Stable-Wronski value. Nevertheless, it makes sense to take a few simple precautions while reading the datasheets of such panels or while buying ones to avoid taking the initial higher output as rated power.

Thin-film solar panels have lower losses. This means that they perform better in:

- Hot climates and higher temperatures

- Low irradiation conditions, i.e., early in the morning, at sunset and in cloudy weather

- Partial shading conditions

They are also more suitable for mounting on a non-standard infrastructure typical for some facades.

Crystalline (mono- or poly-) PV panels are the most common solar panels for home and business photovoltaic systems. Nowadays crystalline solar panels encompass about 90% of the photovoltaics market share.

In contrast, thin-film solar panels retain about 10% of the market share.

Crystalline panels come in a variety of sizes and shapes. The rectangular shape is the most common one.

Every solar panel has its nominal power rated in 'watts-peak' (Wp) or 'kilowatts-peak' (kW), also known as 'installed WpDC power' or 'watts-peak direct current power'.

Here is a comparison between solar panel efficiencies depending on the area needed to install a solar panel of 1 Wp nominal power:

PV cell material	Panel efficiency	Area needed for 1 kWp
Monocrystalline silicon	13-16%	7 m² (75 sq. feet)
Polycrystalline silicon	12-14%	8 m² (86 sq. feet)
Amorphous silicon	6-7%	15 m² (161 sq. feet)

 **Other important issues to consider when choosing the best solar panel are:**

- Solar panel's manufacturer reputability

- Quality of the manufacturing process – whether the manufacturer controls each stage of the process, i.e., Silicon Material->Wafer->Cell->Solar Panel->Solar Module->Solar panel system

- Materials used in solar panels production

Before choosing the best type solar panel for your application, the following factors should be taken into account:

- The prevailing climate conditions – they influence the temperature's impact on solar panel efficiency and the impact of the irradiance changes on efficiency

- Available mounting area – it determines the conversion efficiency of the chosen solar panel type

- Solar panel warranty conditions

- Solar panel manufacturer's reputation

- Your available budget

- Your plans for future expansion of your solar power system

As you see, there is no apparent 'winner' among the different solar panel types. Furthermore, the winner changes depending on different circumstances.

Every advertised 'important parameter' of a given solar panel type, disregarded in the context of the above mentioned variables, could lead you to a wrong decision.

After all, you don't buy performance parameters; you buy better productivity in terms of kWh/kWp generated power provided at the lowest possible cost.

 **Why is it important to be familiar not only with solar panels but also with the other building blocks of a solar power system?**

The efficiency of solar panels to convert sunlight into electricity plays an important role in delivering solar generated electricity.

To produce the electricity needed by your appliances, a solar panel system needs not only solar panels but also certain additional components. Using these additional components introduces unavoidable additional losses in the system, which in turn reduces the amount of electricity produced by the solar panels.

By carefully designing the solar panel system and skillfully selecting all of its building blocks, such losses can be minimized. Moreover, upon everything else being equal, by using special buildings blocks, you could squeeze up to 30% more power from the solar panel system while maintaining its initial price point.

What is more, there are different types of solar panel systems, depending on whether a solar panel system is connected to the utility grid or not. Those different types of solar panel systems include a number of different components.

Mixing solar panels – Dos and Don'ts

Whether it is okay to mix solar panels of different voltage or wattage, or produced by different manufacturers, is a frequently asked question by many DIYers.

Though mixing different solar panels is not recommended, it's not forbidden. It can be OK, as long as each panel's electrical parameters (voltage, wattage, and amps) are carefully considered.

When considering whether to wire two panels produced by two different vendors, know that the problem is not the different vendors. The problem is the different electrical characteristics of the panels and the varying levels of performance degradation.

Solar modules are connected in a series in order to obtain higher output voltage. The maximum system voltage, however, must not be exceeded.

For modules connected in a series, the total power is calculated as follows:

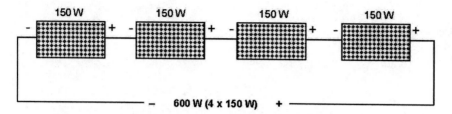

Total connected power = 150W + 150W + 150W + 150W = 600W

However, if, among the modules connected in a series, a module has a rated power lower than the rated power of the other modules, such a module will drag the overall system output down:

Solar modules are connected in parallel in order to obtain higher output current.

For PV modules connected in parallel, the total power is calculated as follows:

Total connected power = 150W + 150W + 150W + 150W = 600W

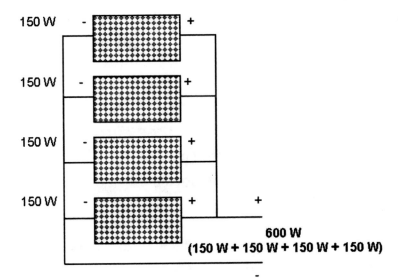

Unlike connections in a series, if among the connected parallel modules there is a module of power output lower than the output of the other modules, this might not seriously affect the total power output of the array, provided that such a module is of the same voltage as other ones:

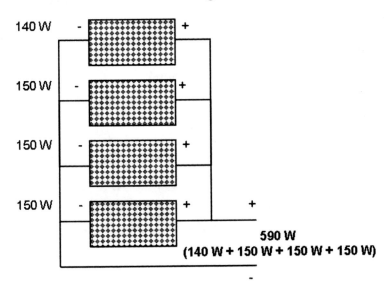

The maximum voltage on a string of modules must always be lower than the maximum input DC voltage of the inverter.

When connecting different solar modules, it's not different wattages, it's actually the current (for series connection) and voltage (for parallel connection) that could drag down the performance of the whole array.

Only solar panels of exact or similar current should be wired together in a series. When you connect a 3A panel to a 3.5A panel, the overall current will be dragged down to 3A. Such a reduction in current will lead to a reduction in power output, and therefore, to a loss in system performance.

Similarly, only solar panels of exact or similar voltage should be wired together in parallel. When you connect a 15V panel to a 24V panel, the overall voltage will be dragged down to 15 Volts. Such a reduction in voltage will lead to a reduction in power output, and therefore, to a loss in system performance.

Compared to voltage and current, wattage is not such a significant concern. When you wire together a 60W panel to a 100W panel in a series, the total connected power would be 160W, provided the two panels are of equal current.

Here the difference in voltages is not important, as voltages will just sum up, and all you have to be careful about is making sure that the total voltage falls within the voltage window of the inverter.

If current ratings, however, are different, you can expect unpleasant surprises since the overall current would be the lower of the two ones, which means that you're not going to obtain a total of 160W, but instead will always achieve less. How much less – it depends on the difference between the rated currents.

Furthermore, when you wire a 60W panel together to a 100W panel in parallel, the total connected power will be 160W, provided that the two panels are of equal voltage. Here the difference in currents is not important, as the currents will just sum up, and all you have to be careful about is making sure that the total current does not exceed the maximum input current of the inverter.

If, however, voltage ratings are different, you can expect unpleasant surprises since the overall voltage would be the lower of the two ones, which will mean that you're not going to obtain a total of 160W, but instead will always achieve less. How much less – it depends on the difference between the rated voltages.

Why you should not connect different solar panels:

1) Apart from rated power, each panel has its power degrade percentage. Therefore, you should expect different output degrade of any solar panel over time.

   Moreover, the degradation stated in the technical specs does not always coincide with what is stated on a panel's nameplate. Therefore, it is not easy to find an exact panel match from different solar vendors. 'Exact match' means both similar rating and similar rating degradation.

   For panels connected in a series, voltage is additive while current is the same, provided all the panels are of equal current rating. If, among the panels connected in a series, there is a panel with a rated current lower than the rated current of the others, it will drag down the current that is passing through all of the remaining panels.

   Therefore, each of the remaining panels (of a higher current rating) will underperform: it will produce current (and therefore power) lower than the current stated on its nameplate.

In other words, if two dissimilar modules are wired in a series, the voltage is still additive, but the current will be only slightly greater than the current produced by the panel with the lowest current output in the series string.

2) For panels connected in parallel, the current is additive while the voltage is the same. If there is a panel among the panels connected in parallel with rated voltage lower than the rated voltage of the others, it will drag down the voltage of all the remaining panels. Therefore, each of the remaining panels (of a higher voltage rating) will underperform: it will produce a voltage (and therefore power) that is lower than the voltage stated on its nameplate.

3) Mixing solar panels with different electrical characteristics is not recommended if you use an MPPT charge controller. Different wattages make it impossible for the controller to find the optimal operating voltage and current since they are different for each panel type.

The solution is simple: utilize panels that have electrical characteristics similar to the original panels.

In order to minimize the losses when connecting different solar panels:

- Connect in series only panels of the same brand and manufacturing series.

- Connect in a series only panels of different brands and of the same current – this is your second option, if you cannot find panels of the same brand for any reason.

- Connect in parallel panels of the same brand/manufacturing series and of the same voltage.

- Connect in parallel panels of different brands and of the same voltage – this is your second option, if you cannot find panels of the same brand for whatever reason.

- Connecting different solar panels with the same array is not recommended, since as a result the voltage or the current might be reduced. Therefore, if you are planning to use dissimilar panels, try to use ones of comparable stated voltage and current.

- *As a last resort, if you, for whatever reason, have to use solar panels of different series from the same manufacturer or from a different manufacturer, please consider using a separate charge controller or inverter, depending on the solar panel system type you are installing. The higher yield of the solar panels will justify quite quickly all the money spent on the additional hardware.*

Alternating Current (AC) Solar Panels

What are AC solar panels?

Alternating current (AC) solar panels have recently created a lot of buzz among solar enthusiasts. Let's take a look at them and see their advantages and drawbacks compared to traditional DC solar panels.

As you already know, a solar panel generates DC power only. An AC solar panel is nothing more than a solar panel integrated with a microinverter. The inverter converts DC electricity to AC electricity. Usually the microinverter and the pertaining cables are attached to the solar panel already in factory conditions.

According to the US National Electric Code, an AC solar panel is "a complete, environmentally protected unit consisting of solar cells, optics, inverter and other components, exclusive of tracker, designed to generate AC power when exposed to sunlight."

Advantages of AC solar panels:

AC solar panels produce maximum power in any conditions, thus ensuring:

- Each panel working independently, without negatively influencing the rest panels in the array, if operating conditions are not optimal

- Optimal energy generation when solar array is shaded

- Part of the panels in a solar array to collect solar energy from non-optimal directions different from True South/ North, without decreasing the overall performance of the array, as it is in a solar power system with a central inverter

- Optimal maximum power point tracking per panel, compared to the less efficient tracking of whole array

- Elimination of power losses from cables connecting solar panels in array that are typical for solar panel systems with a central inverter

- Energy production monitoring at panel level

- Simplified system assembly, thanks to the plug-and-play architecture

- Lower installation cost

Disadvantages of AC solar panels:

The life expectancy of an AC solar panel is limited by the lifespan of the integrated microinverter.

In a typical solar power system, the life expectancy of the inverter is 12-15 years. In contrast, the average lifespan of a panel is 25-30 years. Yet there are 40 years old solar panels that are still completely operational. Usually, manufacturers of AC solar panels guarantee that the embedded microinverter will serve you 25-30 years. This, however, is just a guarantee!

The main drawback of solar panel systems based on AC solar panels is their higher final cost, as compared to a system initially designed with a central inverter. Although the price of a single AC solar panel looks enticingly low, after scaling up your system by adding more AC solar panels, you will find out that your system becomes more expensive, compared to a system of the same size with a central inverter.

Therefore, before deciding to use AC solar panels, you should evaluate whether the higher price of such panels would justify the expected energy production increase of your solar panel system and the ensured monitoring and control at the panel level.

Cheap And Secondhand Solar Panels

In this chapter, you will find:

> ➤ *The essentials about cheap and secondhand solar panels*

> ➤ *How to perform an assessment of secondhand solar panels on your own*

> ➤ *Where to buy cheap solar panels*

> ➤ *The most important things to know about do-it-yourself (DIY) solar panels*

> ➤ *How to evaluate a solar electric system based on DIY panels*

Why are cheap solar panels so popular?

In recent years, solar panels have significantly dropped in price. They are now at a point where, "solar PV is now cheaper than oil and Asian LNG (liquefied natural gas)" [1].

What is more, "the cost of solar PV modules could fall beyond most expectations in coming years – and reach a cost of just 25c a watt by 2020" [2].

The price of solar panels, however, remains high, especially if your daily electricity consumption is high.

Buying secondhand or cheap solar panels often appears to be an attractive option.

Is it, however, really beneficial to you?

Why do people buy secondhand solar panels?

Why would someone make solar panels himself, rather than buy them?

This book is written to provide answers to these questions, as well as to disprove some common myths about cheap solar panels.

References:

1.   Solar's Insane Cost Drop | CleanTechnica [Internet]. [cited 2014 Nov 8]. Available from: http://cleantechnica.com/2014/04/16/solars-dramatic-cost-fall-may-herald-energy-price-deflation/

2.   Citigroup: How solar module prices could fall to 25c/watt : Renew Economy [Internet]. [cited 2014 Nov 8]. Available from:

http://reneweconomy.com.au/2013/citigroup-how-solar-module-prices-could-fall-to-25cwatt-41384

Cheap and secondhand panels: the essentials

Decreased lifespan

The required lifetime of the solar panels you will use in your solar panel system depends on the expected lifespan of the system. For example, if you need a small and cheap solar panel system to provide power to a simple device or a recreational vehicle, a cheap panel is the better option, rather than buying an expensive branded solar panel.

If you buy a cheap solar panel, you are likely to save up to 50% off the price, compared to buying a branded panel. Even if in the beginning the stated parameters turn out to coincide with the real parameters, you could expect some degradation in performance. The worst, however, is that no one can tell you **how** much **further** panel efficiency is going to degrade. Therefore, you might encounter some unpleasant surprises.

So if you choose to buy cheap solar panels, you should buy more powerful ones (that is, of higher rated power each) than you need. A recommended percentage is 15 to 20% higher rated power. Certainly 15-20% more rated power means a higher price, but you still save money and, what is more important, this is a fair guarantee that your system will not underperform years after installation.

The warranty of cheap solar panels is usually only extended to one or two years, rather than the typical 25-30 years warranty standard with new panels.

If you're buying a large-area panel or a solar panel that is to be installed on a vehicle, you should choose a panel made of tempered glass. Tempered glass is 7-8 times more robust than plate glass and is better protected against both physical damage and water infiltration.

Degrade in efficiency

Secondhand panels usually have much lower efficiency. This means not only that you cannot make a precise enough sizing of your solar panel system, but also that you need a greater (how much 'greater' no one can tell!) space to install the modules to generate the same amount of power, compared to the case of using new panels.

 **Warehouse-stored secondhand solar panels do not degrade in efficiency over the years, compared to panels exposed to direct sunlight.**

During the panel survey, you should ask about panel's initial efficiency and how much the initially rated power got reduced.

If the initial efficiency is said to have been reduced by 20%, you should multiply the area required for solar panel installation by 1.2. Therefore, to keep the system output equal to the initially planned one, you have to increase the installed rated Wp power by 20%.

This means adding more panels of the same type. If the initial efficiency is said to be reduced by 10%, you should multiply the calculated solar panel area by 1.1, thus adding more panels to compensate for the reduced efficiency.

 **Remember the golden rule about solar panel assembling:**

Mix neither solar panels produced by different vendors, nor solar panels produced by the same vendor but of different characteristics, or panels belonging to different series!

Branded vs. unbranded secondhand panels

Certainly, not all solar panels are designed to be equal. It is always better to buy a branded panel than an unbranded one, bearing in mind that unbranded panels might turn out to behave in a way other than expected – for example, resulting in a shorter lifespan and/or less energy generated.

If you are implementing a solar panel system that is to operate for a long enough period – 20 years or more – you should choose branded panels for it.

If secondhand panels come from a well-known brand, they are often a really good value for the price. A branded secondhand panel might be able to provide service for 25 years or more.

This is because no one actually knows how much longer the life of a branded solar panel could last. Practice has shown that panels manufactured 30 years ago have a real power output of up to 90% of the original rating.

Regarding the type of solar panels, it is recommended to choose monocrystalline ones, as this technology has been proven over time, and you can still find solar panel systems equipped with these panels working between 25-40 years, with performance degradation of about 20%.

The second best option is to choose branded polycrystalline solar panels. But since this is a relatively newer technology compared to monocrystalline one, nobody can predict their performance degradation over time.

Currently, it is not advisable to buy secondhand thin-film (amorphous) panels since this technology is relatively new, and there is not enough statistical data on the degradation of their performance over time.

What should a secondhand solar panel look like?

A solar panel that has a chipped or broken surface, as well as a solar cell peeling from the glass, is worthless. By buying a worthless solar panel, you are at a loss, no matter how cheap the panel might be.

 **Never buy secondhand solar panels without having a look at them!**

The first thing you are going to notice will be some kind of obvious damage – cracks, scratches, etc.

Above all, you should check whether the panel surface is chipped or broken. You should also watch out for cell damage, which is a typical consequence of improper handling during mounting, unmounting or transportation.

Broken terminal connections are not terribly important, as they can be fixed by soldering. Check whether each connection is fixable – this is one of the reasons that you should bring a fellow electrician or technician with you.

You should also check for water condensation between the glass and the silicon. Moisture has a detrimental impact on the panel's efficiency.

Check for any indications of delamination. Delamination describes the detachment of solar panel layers because of humidity entering into those layers, and appears in the form of bubbles in the panel.

Not all solar panels are resistant to long-term exposure to humidity, so you could ask whether the

Solar panel measurements

You should test the panels using a multi-meter not only to determine whether a panel is operational upon exposure to sunlight, but also to measure the generated voltage.

1) Measure the voltage across the panel terminals. This should read a steady voltage. For example, when exposed to maximum sunlight, a 12-volt panel should show a fairly steady (not fluctuating!) 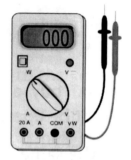 reading of 21 volts between its terminal connections. Then you should measure the voltage across the panels at less sunlight. In total darkness, the voltage should drop significantly, but there should still be some voltage. In shade, the voltage should drop only a little.

2) Examine the current. Change the multi-meter to 'DC current' mode and select the highest amperage scale. Similar to voltage measurement, examine the current across the panel in clear sunlight, in total darkness and in shade. The current could vary a lot more than voltage does. What you are measuring is

the short circuit current of solar panels since the load resistance of multi-meter switched to a current measurement mode is almost zero.

Different panels behave differently. For polycrystalline panels, the cells are connected in a series, so if just one cell is shaded, the current drops significantly. This means that any potential shade is detrimental. Amorphous panels do not have this same problem.

What we are interested in is solar panel output. Power is dependent on the amount of sunlight. To get higher power, you need more sunlight.

Warranty and warranty support

Usually, secondhand solar panels do not come with a warranty. New cheap solar panels, however, normally come with some warranty, although it's much less than 25 years.

Therefore, you should check to see how the warranty would be applied.

This means that you need to ask who will take care of the panel in the event of a problem (malfunction or damage) – a vendor's representative or you personally.

Furthermore, if a panel gets damaged, it is important to know:

- Whether the vendor is going to bear the expense of transportation and repair, and

- In what time they can do that.

Remember that, regarding warranty issues, big solar suppliers are more reliable.

Where to buy secondhand panels?

A good deal might be buying discarded or used solar panels from a manufacturer.

Usually such panels are cheaper due to loose or disconnected wirings (which is not a problem, as they can be easily soldered), while efficiency is almost the same as in new panels.

Another place to get secondhand solar panels is eBay or Amazon. We recommend buying secondhand panels produced by a reputable manufacturer.

Here are some other ideas of where to get secondhand and cheap solar panels:

- Local solar installers – they can advise you where to search them.

- Building contractors specialized in house renovation – they often remove panels from houses they work on, and you can get such panels free of charge if you take part in removing them from a site.

- Insurance companies – an insurance company has a lot to do in areas after a windstorm, a hailstorm or a hurricane, as there are lots of damaged houses and, therefore, damaged solar panels. Some of those panels can be easily fixed, and they might be useful in a DIY solar panel system.

- Secondhand electrical shops

- Newspaper ads

 What to remember about secondhand solar panels

- Having your solar panel system built with secondhand panels, you might not be eligible for any grants, rebates or incentives.

- You will not be allowed to have your solar panel system connected to the grid.

- If panels are not provided with a certificate, this will be a problem for your insurance company.

- Do not buy secondhand panels before having a look at them. During the survey ask: where they were used, what their age is and what is the efficiency degrade.

- Buy secondhand panels only when you're either on a very tight budget or when you need them for small electrical appliances (laptops, lamps, fans, water pumps).

- Secondhand solar panels always require a larger area to install.

- Avoid buying thin-film secondhand panels, try to find crystalline (mono- or poly-) ones.

- Filter out scams – some 'vendors' might even try to sell you apparently promising solar panels manufactured by technology no one has ever seen!

 What to do before buying secondhand solar panels

- Insist on having a look at the panels – arrange a survey.

- Take a fellow electrician, technician or a technology-skilled person with you, and ask him to bring a multi-meter.

- Carefully examine each panel and be careful about breaks, chips, scratches, water condensation (see above).

- Have the voltage, power, and current measured (see above).

 What to ask during the 'solar panel survey'?

- How old are the panels?

- In what kind of environment were they mounted? (the best case is to have been stored in a warehouse)

- How much is the efficiency degradation?

- Do panels come with any warranty and how long is it?

Anyway, is it worth using secondhand solar panels?

Using secondhand solar panels is not advisable, unless either you are on an extremely limited budget or you need to power a small appliance rather than an entire household.

It's a good idea to evaluate first what rebates and incentives you are entitled to in your area. In some cases, such benefits might drag down the total system cost by 30%!

Therefore, you should choose the preferable option for you:

- To have a brand new solar panel system, with warranty and post warranty support, provided with insurance, at no risk, no unpleasant surprises, with at least 25 years lifespan, at a price reduced by 30% (certainly in the case that you are eligible for solar rebates/incentives) from the initially stated one, or

- To have a system built of cheap solar panels – well, probably cheaper than 30% or more but... lacking any warranty support, insurance, always prone to risks and hazards, of unknown lifespan and most importantly – not eligible to benefit from any rebates or incentives?

Mind you, an often forgotten yet proven way to reduce the price of your solar panel system – reducing your electricity needs. Please, don't laugh! The less electricity used, the cheaper the solar panel system needed – inclusive of components, installation, commissioning, insurance and maintenance.

If you buy secondhand or cheap solar panels, your system cannot be insured under a guarantee, so if a panel breaks down or is damaged, you will both have your money lost and your system non-operational. If you care about warranty support, then buying secondhand or cheap solar panels is not an option for you.

Moreover, by using secondhand or cheap solar panels, you could expect possible failures and breakdowns, as well as occasional time-consuming repair work.

Are DIY solar panels an option?

Do-it-yourself (DIY) solar panels are said to be the lowest cost option for installing solar panels.

Making solar panels at home is simply not as easy as described in books and articles. While you may have watched YouTube videos and read all sorts of how-to guides, the construction of solar panels does require a certain level of skill to create a high quality, effective, and safe product.

Homemade solar panels are NOT recommended for use in systems that require high wattage/voltage/current, including those for powering an entire household.

Here are the reasons:

1) Shorter lifespan and much faster efficiency degradation than manufactured solar panels.

Unless you are able to encapsulate your solar cells effectively, water will infiltrate the product and over time, the panel will degrade. This means that your DIY solar panel could have a lifespan of only a few years. Just compare this to the common 25-year lifespan of factory-made panels, combined with a 25-year performance guarantee!

What is more, the possibility of a hailstorm damaging a branded solar panel is rather low. This, however, is not true for a DIY solar panel. Manufactured panels have been designed to withstand the various elements whether it is hail or snow – DIY products simply do not have the same level of detail.

2) Could be a fire hazard resulting from poor quality soldering, especially when combined with the high voltages that occur when connecting several panels together in a string. Many of the DIY solar panel systems fail to address safety. Homemade solar panels should not be made using wood and/or plastic due to the potential for electrical short and resulting fire hazard.

3) Lack of proper certifications. This means that:

- You cannot connect your house to the grid, since homemade solar panels are not compliant with the applicable electrical codes.

- You cannot apply for a government rebate or incentive.

- If a DIY solar panel ignites, and this results to fire damage to your house, your insurance company will not pay up an indemnity for a fire caused by a solar panel with no UL certification (for the US).

- Furthermore, since DIY panels are not certified, mounting them on any insured structure will void the insurance policy on that structure.

 **DIY panels are NOT a viable alternative to factory-made solar panels when it comes to:**

- Having a safe and reliable solar electric system

- Cutting your electricity bills

- Qualifying for rebates and government incentives

There are countless green websites and DIY books full of stuff that seems to work. The question is, how long will such a solar panel system last when exposed to the weather?

If you've never done DIY before and you're now taking your first exciting steps in photovoltaics, solar panel assembly is not a great place to learn electrical wiring and soldering.

Above all, you need some real practice, rather than starting out by damaging your solar cells with an idea to make a cost-effective, working solution!

Evaluating a solar panel system based on DIY solar panels

On the next pages, you will learn how to estimate whether it is worth investing in DIY solar panels and a DIY solar panel system.

We are going to calculate:

- The price of building the DIY solar panels yourself

- Overall system cost including the solar panels and the rest of the solar panel system building blocks needed

- Annual maintenance cost of your solar panel system.

Let's consider the following example of a PV system built of DIY solar panels.

The solar panel system should provide 200Wh daily for area of PSH (Peak Sun Hours)=3.

Peak Sun Hours (PSH) is a measure of the available daily solar resource. **PSH** are also known as 'insolation' and **depend on the specific geographic location.**

Please, do not mistake Peak Sun Hours with available sunny hours!

For example, 6 available sunny hours on a bright sunny day might only translate into 4 PSH for that day.

The system is comprised of the following components:

- Solar panel

- Charge controller

- Battery

- Wires to connect the solar panel to the charge controller, the battery and the loads.

The system voltage is 12V.

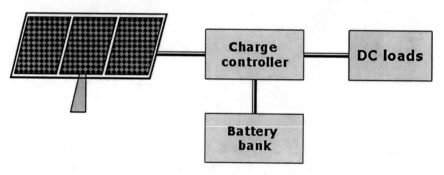

A solar electric system with DC loads

We are interested in and we are going to calculate the following key values:

- Installed peak power required, Wp

- Minimum capacity of the battery bank required, Ah.

The installed peak power is calculated as follows [1] [2]:

Installed peak power, W =

= Daily energy consumption, Wh ÷ PSH ÷ System efficiency

Where:

- Daily energy consumption is the daily averaged or daily peak electricity consumption in Wh,

- PSH is the Peak Sunny Hours value at the system location in hours,

- System efficiency is the efficiency of the solar panel system.

For a stand-alone PV system, the system efficiency is usually evaluated at about 0.6-0.67. If the daily energy target is 200Wh and PSH=3 (see above), then:

Installed peak power, Wp = 200Wh ÷ 3 ÷ 0.67 = 100Wp

The minimum required battery bank capacity is [1] [2]:

Battery bank minimum capacity, Ah =

= (Daily energy consumption, Wh x Days of Autonomy) ÷ Depth of Discharge ÷ Cable Losses correction factor ÷ System voltage, V

Where:

- Daily energy consumption is the daily averaged or daily peak electricity consumption in Wh;

- Days of Autonomy (DoA) is the desired number of consecutive days that we would like the battery bank to power the load in case of a complete lack of sunshine;

- Depth of Discharge (DoD) defines up to how much percentage the battery bank should be discharged, i.e., 100%=empty battery bank and 0%=full battery bank); in the formula the decimal value rather than the percentage should be used;

- Cable Losses correction Factor is the overall system losses due to cable resistance in %; in the formula, the decimal value rather than the percentage should be used.

So, if we assume:

➢ Days of autonomy = 3

➢ Depth of discharge = 0.8

➢ Cable Losses correction factor = 0.97,

then for daily energy consumption 200Wh and system voltage 12V, the minimum required battery bank capacity would be:

Battery bank minimum required capacity =

$$= (200Wh * 3) \div 0.8 \div 0.97 \div 12V = 64Ah$$

So the system should comprise: a 100Wp solar panel for system voltage of 12V, a charge controller and a 64-Ah battery bank of 12V.

In addition, we need a 7m-long wire of a 4-mm2 cross section for connecting the solar panel to the charge controller.

The next step: what kind of charge controller to choose?

Since we are building a low-power low-cost system, we are going to use a PWM controller.

We have a 100Wp solar panel for a 12V solar panel system. Therefore, we need a PWM controller with rated current of

$$100 \text{ W} \div 12 \text{ V} = 8.33 \text{ A}$$

We should enter a safety margin of 25% to account for the changing solar panel output depending on the temperature. Therefore, we need a 12V PWM controller with rated current not less than

1.25 * 8.33 A = 10.4 A

If you want to conform to the National Electric Code (NEC) standard, you have to "de-rate" once again the received value by 25%. Therefore, the current should be modified as follows:

1.25 * 10.4 = 13A

You should, however, take into account that some of the established brands, e.g., Morningstar, give for a PWM controller's rated current the de-rated value, which accounts for a 25% additional increase.

Please, carefully check this in advance because in such a case you should only multiply the rated current by 1.25, in order to adhere to the NEC standard applicable for the USA.

Furthermore, the standard recommends performing such calculations by using the solar panel's short circuit current (Isc), rather than by using the presupposed working current we've calculated in our example so far.

For such a purpose, you should have a solar panel's datasheet at your disposal. If you do not have such a datasheet available, you could use the method described above to determine the charge controller current, by dividing the solar panel rated wattage to the system voltage.

Also, don't forget that in case of several solar panels being connected in parallel into a solar array, the resulting current of the array is a sum of the currents provided by each solar panel.

So, if you consider our example, in a case of such 'already de-rated' charge controller, its current rating must be greater or equal to 10A.

Why do you need two de-rating coefficients of 25%?

The first de-rating coefficient ensures that the charge controller will withstand any instantaneous changes in the solar panel's current because of constantly changing ambient temperature and solar irradiation.

The second de-rating coefficient ensures that the charge controller will withstand the continuous load of the increased solar panel current for three hours or longer. Therefore, you have to multiply the solar panel's current by 1.25*1.25=1.56.

Don't forget, however, that in the case of an already de-rated branded charge controller, you should only multiply the solar panels current by 1.25, in order to comply with the NEC standard. What is more, in such calculations it is recommended to use the short circuit current (Isc) of the solar panel or the solar array, instead of forecasted solar panel's nominal operating current.

Now, let's summarize. In our example, if you choose to use a brand charge controller whose current has already been "de-rated," you need a PWM charge controller with rated current greater or equal to 10A. Otherwise, you need a charge controller with rated current greater than or equal to 13A.

The price of such a branded PWM controller is about $40. Actually, on eBay, you can find a cheaper PWM controller from a less reputable brand for around $15, but this will be at the expense of non-guaranteed performance characteristics and eventually shorter battery life. The choice is yours.

Now, let's calculate the materials needed for building the solar panel.

If you are aiming at the lowest cost solution, you have to build your panel out of polycrystalline cells.

One polycrystalline cell typically gives 0.56V and about 6 to 8 Amps of current, depending on the cell grade/quality. Mind that the higher the quality, the higher the current.

Therefore, if you connect 36 cells in a series, you will receive a panel of voltage about 20V and current of about 6A, which translates into about 120Wp of power. The 20V panel output will guarantee that under all operating conditions, the panel output voltage will be high enough to charge the battery bank.

To assemble that solar panel we need:

- 36 polycrystalline cells – **$40** (on eBay)

- Wiring kit – **$25** (includes a tab wire, a bus wire, a flux pen, a junction box with Schottky diodes) – sometimes is provided as a bonus, if you are lucky.

- Clear Plexiglas or low iron solar glass – **$40**

- Plywood panel – **$25**

- Paint, stainless screws, wood for frame – **$10**

- Your dollar time value to assemble this panel (this is your hourly pay rate from your employer) – **$X.**

Total cost = **$140**, excluding your dollar time value, plus the price of additional cells you need in case of breaking some cells while assembling them.

If you are inexperienced, you'll inevitably break some of them – don't fool yourself!

You can, however, find similar first-hand panels for $160-$180 with 25 years of warranty. This information is just for your records.

In summary, here is what we have calculated so far:

Therefore, the total price of the system is $3.9 per installed Wp.

Item	Price
DIY Solar Panel 100Wp	$140
12V/10A PWM charge controller from an established brand	$40
Solar Battery 12V/64A	$200
Wire 4 mm² for solar $2 per meter	$14
Total	$394

Let's assume a lifecycle of 25 years of operation. We are going to calculate the system's annual maintenance costs [2]. The system contains a battery and a charge controller, so we need to calculate:

➢ The battery's annual maintenance costs

➢ The charge controller's annual maintenance costs.

A lead-acid battery has to be replaced every 5 years of operation. Let's assume that the battery bank cost, stated by the solar vendor, is $3 per Ah. In addition, the battery cost for the first 5 years is included in the system's cost.

If the stated battery cost were $3, then for the next 20 years of the system lifecycle, the costs for a battery bank of 64 Ah would be:

64 Ah * (20 years ÷ 5) * $3/Ah = $768

Such a cost, distributed over the remaining 20 years of operation, will result in average battery maintenance costs per year as follows:

$768 ÷ 20 years = $38.4/year

Furthermore, for a PWM charge controller of $40, which is not likely to be replaced during these 25 years, the annual maintenance costs would be:

$40 ÷ 25 years = $1.6/year

Therefore, the total annual maintenance costs would be [2]:

$38.4/year + $1.6/year = $40/year

Consequently, the estimates show that annual maintenance cost of this system is about $40, upon assumed 25 years of operation.

References:

1. Mayfield, Ryan. 2010. Photovoltaic Design and Installation for Dummies, Wiley Publishing Inc.

2. Antony, Falk, Christian Durschner, Karl-Heinz Remmers. 2007. Photovoltaics for

Professionals: Solar Electric Systems Marketing, Design and Installation, Routledge.

3. Endecon Engineering. 2008. A Guide to Photovoltaic (PV) System Design and Installation, June 14, 2001.

4. German Energy Society. 2008. Planning and Installing Photovoltaic Systems – a Guide for Installer, Designers and Engineers.

5. Munro, Khanti. 2010. Designing a Stand-Alone PV System, Home Power Magazine 136, April-May 2010, pp.78-84.

6. Solar Energy International. 2007. Photovoltaics: Design & Installation Manual, New Society Publishers.

How to Compare Solar Panels to Choose the Most Efficient, Reliable and Cost Effective Ones

 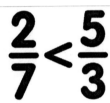 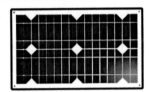

$$\frac{2}{7} < \frac{5}{3}$$

In this chapter, you will find:

➤ *How to properly grasp a solar panel datasheet – learn how to read 'between the lines'*

➤ *Why selecting the 'right' solar panel can save you a lot of money*

➤ *Some tricks used by solar manufacturers*

➤ *How to compare solar panels produced by different manufacturers*

Solar panel datasheet demystified

A solar panel datasheet contains the main specifications and key features of a photovoltaic module.

This is the data you need to integrate solar modules into a solar power system.

'Nominal power' is what is stated as power in Wp (watts-peak) in the solar panel datasheet. Therefore, 'nominal power' is what you get as the power output from a photovoltaic panel. Nominal power is important for solar panel system sizing and evaluation.

As ambient temperature increases, solar panel efficiency (the ability of a solar panel to convert solar energy into electricity) gradually decreases. Therefore, at a certain high temperature, the generated solar electricity reaches its lowest value.

Furthermore, as temperature decreases, efficiency increases.

This means that if you have bought a solar panel of 200Wp nominal power, it might generate 20-30% less electricity in summer than in winter.

For various panels, such a temperature dependence of the solar generated electricity will vary. It might surprisingly turn out that in your climate, a more expensive solar panel will produce much more electricity annually than a less expensive one, provided that both panels are of the same nominal power. Thus, a more expensive panel could definitely turn out to be more beneficial from a financial point of view.

Cost per watt, or $/W, is a commonly adopted way to compare costs of generating electricity. For photovoltaics this refers to the solar generated electricity and is calculated by dividing the total investment costs of a solar project by the total power of the solar panels installed, also known as 'peak power' (watts-peak or 'Wp').

 Example:

A solar panel of 220 Wp nominal power, bought for $270, will have a cost-per-watt value estimated as follows:

$$\$270 \div 220 \text{ Wp} = 1.23 \text{ \$/Wp}$$

Why the universal measure 'cost per watt' could mislead you?

 Example:

You are faced with the option of either buying a 220Wp solar panel for $270 or buying a 180Wp panel for $221, and you are wondering which option is better.

From the Cost per Watt universal measure point of view, both panels are equal since for both options the Cost per Watt is estimated as equal to $1.23.

It could turn out that, considering the seasonal changes in temperature, the lower rated output panel of 180Wp will generate more energy on an average annual basis than the higher rated power panel of 220Wp.

Furthermore, from the required installation area point of view, it could turn out that a solar panel of a lower output could be more advantageous for you since it could require a smaller area to install.

Be sure that if you use the nominal power data provided by the manufacturer, without even considering the better efficiency of the lower rated panel under the environmental influence, you could easily end up in a situation when you need to install a greater number of panels.

 **Don't forget that a solar panel system's lifecycle is at least 20-25 years!**

If you do not perform such an evaluation, or let yourself be misled by a solar vendor or installer, the money lost as a result of underperforming solar panels or solar panel system, could amount to a couple of thousand dollars across the system's operational period!

The most important solar panel rating or how NOT to lose money as a consequence of misinterpretation

The lines below are quite significant, as they could save you a lot of money.

Here we are going to help you choose the best solar panel in terms of money and guaranteed performance over the years, by understanding key performance ratings of solar panels.

Thanks to this info, you will be able to find the "right" solar panels for your solar endeavor. Furthermore, you will be able to look for important words and parameters in solar panel datasheets that solar manufacturers use to "conceal" inferior solar panels.

By the end of this chapter, you are going to find how the lack of just one word in a solar panel guarantee statement might cost you a lot of money in the long run.

Also, don't forget that the right choice of solar panels is 80% responsible for the overall performance of your solar power system. Get it wrong and you'll lose a lot of money in years to come!

This chapter might seem boring at first glance. But we believe that the eventual financial reward is worth your patience!

So, let's start our journey in the world of a typical solar panel datasheet!

The essential solar panel parameters

1) Solar panel rated power – the DC power produced by a solar panel under the Standard Test Conditions (STC):

- 1,000W/m2 of sunlight intensity

- 25°C (77°F) cell temperature

- Spectrum at air mass of 1.5 [1].

Solar panel rated power is also known as **'nameplate rating'**.

The STC have turned out to be not quite a precise measure mostly due to the requirement to measure the power under factory conditions by maintaining a temperature of 25°C (77°F) of the solar cells.

Due to panel encapsulation, the solar cell temperature is 15-20°C (59-68°F) higher than the ambient temperature. This means that at a cell temperature of 25°C (77°F), the ambient temperature is usually 5-10°C (41-50°F). Obviously, such a testing mode would not completely overlap with the real conditions under which a solar panel typically operates.

As a result, in 1990, as a part of the US project PVUSA (Photovoltaics for Utility Scale Applications), a new rating system was developed for evaluation of solar panel power. Later this system was adopted by the California Energy Commission.

Because of this, on solar panels datasheets you are likely to see the CEC PTC rating that is an abbreviation of 'California Energy Commission's Photovoltaics for Utility Scale Applications' (PV-USA) Test Conditions rating.

The conditions under which the CEC PTC rating is measured are:

- 20°C (68°F) ambient temperature

- Elevation of 10 meters above ground level

- Wind speed of 1 m/s [2].

Since while increasing the ambient temperature, the power generated by the solar panel decreases, the CEC PTC power is always lower than the power measured under the Standard Test Conditions.

When comparing a pair of solar panels, the panel of higher PTC rating is better, since it is quite possible two panels manufactured by two different manufacturers to be of equal STC nameplate rating and of different PTC ratings at the same time.

Later we are going to see that the better way to compare two solar panels is not only by comparing their powers but also by considering the generated PTC or STC power per required installation area (in m2 or ft2).

You can find more information about the STC and PTC power rating of different solar panels here:

http://www.gosolarcalifornia.org/equipment/pv_mo dules.php

2) Solar panel efficiency When we divide this nominal rated power to the sunlight power falling onto a solar panel's area, we obtain the solar panel efficiency. In other words, solar panel efficiency is a measure of the solar panel's ability to convert solar energy to electricity. Certainly, the higher the efficiency, the lower installation area required per generated 1W DC power.

3) Maximum Power Voltage (Vpm) – the solar panel's voltage measured at maximum power (STC).

4) Maximum Power Current (Ipm) – the solar panel's current measured at maximum power (STC).

5) Open Circuit Voltage (Voc) – the voltage of the solar panel measured under STC conditions without a load attached thereto. In out-of-lab conditions, you can obtain an approximate value of Voc by measuring this voltage by using a voltmeter as a load and as a measuring device simultaneously. This means that you should only plug a voltmeter in at the solar panel's output. Certainly, the measured value of the Voc will be closer to the stated one, as the measuring conditions are closer to the STC.

Attention: Such a measurement circuit is only valid when testing a single panel. Testing a solar panel string can result in damaging the measuring device, while the high voltage and current of the string are dangerous to your life.

6) Short circuit current (Isc) – the current of the solar panel measured under STC conditions, when the panel output is short-circuited. In out-of-lab conditions, you can obtain an approximate value of Isc, by measuring this current by using an ammeter of almost zero resistance as a load and as a measuring device simultaneously. Certainly, the measured value of the Isc will be as closer to the stated one, as the measuring conditions are closer to the STC.

Attention: Such a measurement circuit is only valid while testing a single panel. Testing a solar panel string can result in damaging the measuring device, while the high voltage and current of the string are dangerous to your life.

7) Temperature coefficient of the rated power – describes how much solar panel output power deviates from the value measured at the rated power conditions.

If power is rated under the Standard Test Conditions (STC), this coefficient shows how the generated power differs from the power generated at a cell temperature of 25°C (77°F). This coefficient is negative for all types of solar panels, which means that the higher the cell temperature above 25°C (77°F), the lower the power output.

HELP **_Example:_**

If the temperature coefficient of a solar panel at STC is -0.5%/°C, and the ambient temperature is 40°C, the cell temperature is roughly expected to be 15°C higher than the ambient temperature, as a result of panel encapsulation.

The loss in the solar panel power output at 40°C, compared to power output at a cell temperature of 25°C, would be:

$$0.5\%/°C*(55-25) = 15\%.$$

In other words, a panel rated at 100W under the STC (25°C cell temperature), would only produce 85W at 40°C (104°F) ambient temperature.

 **The less negative value of temperature coefficient of the rated power, the better the solar panel.**

The current and voltage of solar panels also have their temperature coefficients listed. Similarly, you should note that the temperature dependence is given in regards to reference cell voltage and current (STC@25°C).

8) Watts per sq. foot / sq. meter – a very important parameter to consider when buying solar power. It shows whether the available installation area would be enough to allow the generation of the solar power you need or want to install.

Below you can find a comparison table showing the required area for producing 1kWp (1,000Wp) STC power for the three types of solar panels [3].

You could use this table to estimate roughly the solar panel technology type that will best match your solar project, provided you already know the necessary STC power to be installed.

PV cell material	Panel efficiency	Area needed for 1 kWp
Monocrystalline silicon	13-16%	7 m² (75 sq. feet)
Polycrystalline silicon	12-14%	8 m² (86 sq. feet)
Amorphous silicon	6-7%	15 m² (161 sq. feet)

9) Warranted tolerance is also known as 'manufacturing tolerance in percentage.' This is a very important parameter often used by manufacturers and solar vendors to trick you.

During the manufacturing process, not all of the panels from a specific model are created equal.

Some of them have the actual rated power equal to the stated rated power; some of them, however, have the actual rated power slowly deviating from the stated rated power.

Therefore, manufacturing tolerance guarantees the range that a solar panel rated power of a certain model might fall within. This value reflects the initial internal stabilization effects taking place in the solar panel as well.

How do solar panels manufacturers and their vendors use this parameter in their marketing favor?

 Example:

Manufacturer X produces solar panels of stated rated power 300W and manufacturing tolerance -5% to +5%.

Manufacturer Y produces solar panels of the same stated rated power 300W and manufacturing tolerance 0% to +10%.

Both panels are marketed as 300W panels and priced equally, let's say, $300. Since both panels are compared based on the Price per Watt ratio, they seem completely equal with a ratio of $1 per W.

Are the above panels really equal?

Will they deliver the same energy for the same price?

Let's check.

If we buy panels produced by manufacturer X, we get panels of actual rated power within the range 285-315W, and a supposed ratio of $0.95-$1.05 per W.

If we buy panels produced by manufacturer Y, we get solar panels of actual rated power within the range 300-330 W and the corresponding ratio of $(0.9-1)/W.

Is 285W solar power equal to 300W? Nope.

Is 285W equal to 330W? Nope.

Is it fair in regard to you, the Customer? We don't think so!

Moral of the story: While designing your solar panel system, always take as the nominal power the lowest value of the guaranteed power range.

In the above example, for the aforementioned solar panel of stated rated power 300W at STC and tolerance -5% to +5%, the real nameplate power during the design of the system should be considered 285W.

Limited Power Warranty

Just another parameter or the next "marketing trick" used in solar panel datasheets?

A typically offered warranty on solar panels is from 5 to 10 years for workmanship and up to 25 years for performance. However, it is important to know how solar panel performance warranties are structured. Let's shed some more light on this issue.

 Example:

1) You have a panel's rated power Pmax stated as 150W, with a production tolerance of ±5%.

2) Here you can calculate the Initial Warranted Power (IWP). IWP is the typically rated power Pmax, reduced by the manufacturing tolerance in percentage, that is:

$$IWP = Pmax - 5\% = 142.5W$$

3) Then you will find that the panel has a 90% warranted stated power for the first 10 years, which means that it is actually 90% of initially warranted power or 128.25W. In other words, only if the solar rated power falls below 128.25W for the first 10 years, you have the right to request a replacement.

4) Then you have an 80% guarantee of the output power stated for 25 years, which is actually 80% of IWP, or

$$0.8 * 142.5 = 114W$$

Similarly, if the solar panel rated power falls below 114W under the above mentioned conditions, you have the right to request a replacement.

So far, so good. Somehow, we could accept these surprises.

This is, however, not the end of the story. What you should look for is whether the money saving term **'linear performance guarantee'** is used.

What does **linear performance guarantee** mean?

Example:

If a manufacturer has stated that a solar panel has a linear performance guarantee of 90% for the first 10 years, this means that the panel performance will degrade by the same value year after year within that 10-year period – or 1% per year. This is a very good panel!

What if you can't find the "magic" word **linear performance guarantee** in the datasheet, and the manufacturer has just stated that the panel has only performance guarantee of 90% for the first 10 years?

This could mean that at the end of the first year of operation the performance of the panel might drop significantly to 91% of its guaranteed power, and during the next 9 years, the performance might only drop off 1% until it reaches the guaranteed 90%.

Have you noticed the big difference? In such a case, for the last 9 years you will incur higher cumulative losses compared to the scenario with linear power guarantee for 10 years.

Although at first glance there is no difference between the warranty statement with a linear and non-linear guarantee, it is not true.

 **Be careful while reading solar panel datasheets!**

Since most solar modules have a limited power warranty, manufacturers are not likely to offer module replacement.

Instead, they would suggest:

- Repairing the faulty PV modules, or

- Compensating the reduced power output by offering enough new solar modules, or

- Refunding you for the decrease in power output by considering the accumulated years of use. For a warranty of 25 years, the annual deduction is typically 4%.

Mind that normally the manufacturer will not reimburse your costs for dismantling, transportation or reinstallation of solar modules.

What is more, warranties normally exclude problems resulting from improper installation, any repairs, changes or dismantling performed by non-qualified staff, as well as accidental breakages or abuses, and also lightning strikes and other natural disasters [3].

Nominal Operating Cell Temperature

As mentioned above, a solar module is usually rated under the Standard Test Conditions that require maintaining the solar cell temperature at 25°C upon measurement.

We also already know that solar cell working temperature is very important for the solar panel's capability to convert solar energy into electricity. The higher the solar cell working temperature, the less the generated solar electricity.

Therefore, in order to squeeze more power from a solar panel, its average cell temperature should be as low as possible. The main drawback of the STC rating, however, is that it doesn't "match perfectly" the real working conditions.

That is why Nominal Operating cell temperature rating was introduced – to give you a more realistic picture of the solar panels performance in real working conditions, by providing you the average value of solar cell temperature of a given panel.

According to the standard, the Nominal Operating Cell Temperature (NOCT) is defined as the temperature reached by open-circuited cells in a solar panel at Irradiance of 800 W/m2 at ambient temperature of 20°C, wind speed of 1 m/s and panel tilt angle of 45°C [4].

What is the average NOCT of the panels currently being sold on the market?

For most solar panels, it is about 48°C (118.4°F). Such panels are considered 'good' ones.

By taking into account NOCT and the temperature power coefficient of a solar panel, we can estimate the power losses of the panel in regards to the reference cell temperature 25°C (STC conditions).

 Example:

If a solar panel has power coefficient of -0.5%/°C and NOCT of 48°C, the expected power performance degradation of the solar panel would be:

0.5% * (48°C -25°C) = 11.5%

or 11.5% lower than the rated power at the STC conditions.

11.5% power degradation is not a low-performance loss!

Low-quality solar panels have NOCT above 55°C (131°F).

 **The moral of the story about NOCT is:**

When comparing two panels, the better one will be the panel of lower NOCT rating!

So be sure to pay attention to this parameter when buying or comparing solar panels.

Solar panel certificates

Why should you only buy panels that are certified?

Over the years, it was observed that after being put into operation, some solar panels showed certain repetitive failures and malfunctions, such as broken interconnects, broken cells, corrosion, damage due to hail, etc. Some of these defects were observed in the beginning, some of them after years of operation.

That is why the International Engineering Commission (IEC) has developed many international standards for solar panel testing. Passing these tests is a guarantee that the chances a certified solar panel will fail when exposed to the conditions for which it was initially tested, are quite low.

Most of those tests are accelerated stress tests. This means that long years of exploitation are simulated by increasing the burden of failure factors over a solar panel for a short period of time.

The corresponding international standards that guarantee the quality of solar modules are:

- IEC 61215 for Crystalline Silicon Modules titled" Crystalline silicon terrestrial photovoltaic (PV) modules - Design qualification and type approval" [5].

- IEC 61646 for Thin Film Modules-titled "Thin-film terrestrial photovoltaic (PV) modules - Design qualification and type approval" [6].

- IEC 62108 for CPV Modules titled "Concentrator photovoltaic (CPV) modules and assemblies – Design qualification and type approval."

According to International Engineering Commission IEC 61215 standard:

"Lays down requirements for the design qualification and type approval of terrestrial photovoltaic modules suitable for long-term operation in general open-air climates, as defined in IEC 60721-2-1. Determines the electrical and thermal characteristics of the module and shows, as far as possible, that the module is capable of withstanding prolonged exposure in certain climates" [5].

IEC 61646 abstract for Thin Film Modules standard reads:

"IEC 61646:2008 lays down requirements for the design qualification and type approval of terrestrial, thin-film photovoltaic modules suitable for long-term operation in general open-air climates as defined in IEC 60721-2-1. This standard applies to all terrestrial flat plate module materials not covered by IEC 61215. The significant technical change with respect to the previous edition concerns the pass/fail criteria" [6].

Another important standard is IEC 61730, which deals with "the fundamental construction requirements for photovoltaic modules in order to provide a safe electrical and mechanical operation during their expected lifetime. Addresses the prevention of electrical shock, fire hazards, and personal injury due to mechanical and environmental stresses" [7].

In many markets, passing the IEC 61215 standard for crystalline solar panels, IEC 61646 standard for thin-film panels and IEC 61730 is considered a minimum requirement to participate [8,9].

 **The best solar panels are the ones that have:**

- Higher CEC PTC rating

- Less negative temperature power coefficient

- Positive warranted power tolerance

- Linear power performance warranty

- Lower NOCT rating

- IEC 61215 & IEC 61730 certificates

- Long-term and well-traced back history of their manufacturers

References

1.   Fraunhofe ISE. STC Measurements — Fraunhofer ISE [Internet]. [cited 2014 Oct 18]. Available   from: http://www.ise.fraunhofer.de/en/service-units/callab-pv-cells-callab-pv-modules/callab-pv-modules/stc-measurements

2.   GO Solar California. California Solar Initiative Program Handbook August 2014 Handbook. 2014 p. 198.

3.   EMA. Handbook for Solar Photovoltaic (PV) Systems. EMA; 2011. p. 64.

4.   Arndt R, Puto IR. Basic Understanding of IEC Standard Testing For Photovoltaic Panels.

5.   IEC. IEC 61215 ed2.0 - Crystalline silicon terrestrial photovoltaic (PV) modules - Design qualification and type approval [Internet]. IEC 61215 ed2.0 - Crystalline silicon terrestrial photovoltaic (PV) modules - Design qualification and type approval. 2005 [cited 2014 Oct 18] p. 93 pages. Available   from: http://webstore.iec.ch/webstore/webstore.nsf/artnum/034077!opendocument

6.   IEC. IEC 61646 ed2.0 - Thin-film terrestrial photovoltaic (PV) modules - Design qualification

and type approval [Internet]. Thin-film terrestrial photovoltaic (PV) modules - Design qualification and type approval. 2008 [cited 2014 Oct 18] p. 81. Available from: http://webstore.iec.ch/webstore/webstore.nsf/ArtN um_PK/39336!openDocument

7.   IEC. Photovoltaic (PV) module safety qualification - Part 1: Requirements for Construction [Internet]. IEC 61730-1 ed1.2 Consol. with am1&2. [cited 2014 Oct 18]. Available from: http://webstore.iec.ch/webstore/webstore.nsf/ArtN um_PK/47647?OpenDocument

8.   John Wohlgemuth, NREL. IEC 61215: What It Is and Isn't (Presentation), NREL (National Renewable Energy Laboratory). 2012;

9.   Paraskevadaki E. The 5 most important things to look for in solar panel data sheet [Internet]. [cited 2014 Oct 12]. Available from: http://www.into-solar.com/2013/12/solar-panel-info-1.html

Photovoltaics In Summary: Pros And Cons

Independence

You gain independence from your utility company.

Environment friendly

Solar energy is clean, sustainable and renewable – unlike the fossil fuel based gas, coal and oil that most utility companies use.

It does not produce sulfur dioxide, nitrogen oxide, mercury or carbon dioxide as byproducts, so it causes no air pollution.

Since it does not emit greenhouse gasses, it does not contribute to global warming.

No noise pollution

PV systems operate silently, with no moving parts.

No fuel costs

No fuel is used, so there are no costs related to buying, transporting or storing fuel.

Low maintenance cost

This is valid for all kinds of PV systems and is not valid for conventionally-fueled electric systems.

Easy to expand

You can add more modules, thus having the ability to generate more electricity.

Highly reliable

Photovoltaics are a proven and reliable technology, especially in tough climatic conditions.

Long life duration

Most solar modules are guaranteed to generate power for 25 years, and they continue to perform well after that time has passed.

Safety

Photovoltaic systems do not work with combustive fuel, so they are safe, as long as they are properly designed and installed.

Add value to property

You can expect the value of your property to increase by 20 times your average utility savings.

Government incentives

Governments offer incentives to people who use alternative forms of energy provided the system being installed complies with government standards.

Prices going down

The price of photovoltaic cells is steadily decreasing, with new types of cost-effective solar modules regularly being developed and introduced to the market.

Worldwide applicable

Photovoltaic panels and systems can be installed almost everywhere. The only requirement is that the site have access to direct sunlight.

Solar electricity might not be economically beneficial for everyone

Although prices of photovoltaics are steadily going down and electricity prices rise gradually in the USA on average with 5% per year, the cost effectiveness of solar panel systems installation depends not only on solar panel prices but also on current electricity price at your location and on hardware installation cost. The cost of a PV system can be further reduced by solar rebates and other incentives.

Therefore, each case is different and needs careful evaluation.

Partially replacing your utility grid with a solar panel system can be beneficial for you depending on your circumstances.

Variability of solar radiation

Unfortunately, the sun does not deliver the same amount of energy at various locations and in various seasons.

In winter, energy generated by a PV system might not be enough to meet daily energy needs. This imposes the use of either a large battery bank or an additional power source, which brings certain drawbacks and higher costs.

Need for specific orientation

To provide maximum energy yield, roofs where solar panel systems are installed should face South (or North, if you live in Australia or New Zealand), and a certain elevation angle is recommended for mounting the PV panels.

Otherwise, the system will not perform well, and the whole investment becomes pointless.

Site must not be shaded

Shade has a negative impact on solar panels and as a consequence, the solar panel system's performance might drop dramatically. Shade and solar panel systems cannot coexist.

While some manufacturers claim that their panels are shade tolerant, do understand that if only 1/4 of the panel cell area is shaded, the generated electrical power will be virtually nil.

Only produce energy in daytime

Solar panels only produce energy during the daytime and the amount of energy provided is different during that period. What is more, a solar panel system can generate excessive energy during given hours when you do not need it.

The solution to such a problem is using an energy storage system (battery bank).

Summary of Links and Resources

Here is the list of websites and solar resources that could be helpful to your solar project.

Free solar panel and solar power calculators:

http://solarpanelsvenue.com/sbfcalculators.php

Top sources of information about solar panels parameters:

A company directory profiling 22,735 solar manufacturers, sellers, and solar panel installers; a product directory presenting 35,648 products' datasheets and pictures

http://www.enfsolar.com/

Large list of solar panels, solar power system building parts and solar auctions:

http://www.renooble.com/

Incentive Eligible Photovoltaic Modules in Compliance with SB1 Guidelines provided by the state of California /USA/

http://www.gosolarcalifornia.ca.gov/equipment/pv_modules.php

Thank you for your interest in our book. We hope you enjoyed reading it.

If you would like to receive more information about solar panels and solar power systems, please visit us at solarpanelsvenue.com.

There you can subscribe to our free newsletter as well.

By doing so, you will receive more information and instant notifications about the new versions of this book and other free content related to solar panels and solar power.

We would be more than happy if our common journey in the realm of solar power continued and we would love to help you even more in your solar endeavor.

Please visit us now at solarpanelsvenue.com.

Last but not least, we'd really love to know what you think of our book. Please, review our book on Amazon. Your opinion does matter! Thank you very much for your time.

Also by the authors:

The Ultimate Solar Power Design Guide: Less Theory More Practice [Kindle Edition]
ASIN: B00Q95UZU0

The New Simple And Practical Solar Component Guide [Kindle Edition]
ASIN: B00TR7IJPU

Top 30 Costly Mistakes Solar Newbies Make: Your Smart Guide to Solar Powered Home and Business [Kindle Edition]
ASIN: B00S8L4IIS

Glossary of Terms

Alternating current (AC) – electrical current changing its direction at a given interval.

Balance of system (BoS) equipment – all the equipment apart from the solar array, which is needed for a solar electric system to operate.

Battery – a device capable to produce DC electricity and store it for later use.

Battery bank – a combination of batteries connected together.

Cable Losses – the overall system losses due to cable resistance.

Capacity of a battery – the amount of electricity a battery can store. Capacity is measured in Ampere-hours (Ah).

Charge controller – a device managing the process of battery charge and discharge.

Current – directional movement of electrons upon certain voltage applied.

Conductor – a stuff where electric current can occur.

Days of Autonomy (DoA) – the desired number of consecutive days that we would like the battery bank to power the load in case of a complete lack of sunshine.

Depth of Discharge (DoD) – defines up to how much percentage the battery bank should be discharged: 100%=empty battery bank; 0%= full battery bank

Disconnect (breaker) – an electric switch protecting an electric circuit from overload.

Direct current (DC) – an electric current flowing always in the same direction.

Distribution panel (distribution board) – a device dividing electrical power supply into several electrical circuits.

Energy – the work that can be done within a certain period of time.

Energy efficiency – a set of measures resulting in electrical consumption reductions.

Fuel generator – a generator working on combustive fuel, able to generate AC electricity.

Grid-tied (grid-direct, grid-connected, grid-on) system – a solar electrical system producing electricity that can be both used in your home/office and exported to the grid.

Hybrid system – an off-grid system that combines photovoltaics with additional power sources (e.g., combustive fuel generator, wind generator, etc.) to produce electricity.

Insulator – a stuff in which no electric current can flow.

Inverter – a device converting DC into AC electricity.

Load – a device consuming electricity to do some useful work.

Net metering – the process of measuring the solar electricity exported to the grid by a solar panel system owner, credited by the local utility company.

Off-grid (autonomous) system – a solar electrical system disconnected from the grid and producing electricity for home/office use only.

Photovoltaic (solar) cell – the smallest semiconductor unit producing electricity when exposed to sunlight.

Photovoltaic (solar) generator – solar array with all its cabling and disconnects

Photovoltaic (solar) module – a combination of solar cells connected together.

Photovoltaic (solar) panel – a combination of solar modules connected together. Often, although not fully correct, terms 'solar panel' and 'solar module' are used interchangeably.

Photovoltaic (solar) array – a combination of solar panels connected together.

Photovoltaic (solar) system – a combination of solar modules and other equipment connected to produce electricity for practical needs.

Power – the rate of consuming/generating energy.

Semiconductor – a stuff where electric current can only occur under certain conditions. Solar panels are made of semiconductor (silicon) material.

Stand-alone system – an off-grid system that uses solely photovoltaics to produce electricity.

Voltage – a difference in the potential (hidden) energy between two points, causing current to flow upon free electrons available.

INDEX

CPSIA information can be obtained
at www.ICGtesting.com
Printed in the USA
FFOW04n0205140116
20440FF